Christian Sekimonyo Shamavu
Marcellin Mulezi

Educating the public about the harmful effects of volcanic ash

Christian Sekimonyo Shamavu
Marcellin Mulezi

Educating the public about the harmful effects of volcanic ash

The case of the Nyiragongo and Nyamulagira volcanoes in the Democratic Republic of Congo

ScienciaScripts

Cover image: www.ingimage.com

This book is a translation from the original published under ISBN 978-3-8416-3644-7.

Publisher:
Sciencia Scripts
is a trademark of
Dodo Books Indian Ocean Ltd. and OmniScriptum S.R.L publishing group

120 High Road, East Finchley, London, N2 9ED, United Kingdom
Str. Armeneasca 28/1, office 1, Chisinau MD-2012, Republic of Moldova, Europe
Printed at: see last page
ISBN: 978-620-7-90440-2

Contents

GENERAL INTRODUCTION

Communities living near active volcanoes may be exposed to respiratory risks from volcanic ash. It is important to understand their perception of the risks and the measures they take to mitigate these risks in order to develop effective communication strategies.

Volcanic eruptions affect the earth and humans around the world. When a volcano erupts and/or when there is volcanic activity in the central crater of a volcano, tonnes of volcanic ash are ejected into the atmosphere and deposited on the ground. The danger posed by volcanic ash is not limited to the area near the volcano, but can also affect a vast area. Ash ejected from the volcano affects people's daily lives, disrupting agricultural activities and damaging crops. **(Dian Fiantis at al. 2019)**

0.0. And the question

- **Espoir BISIMWA KIZUNGU (2013), the** aim of his work was to reduce the effects of volcanic plumes on the health of the population in order to achieve social and health development. The author found that the problem of volcanic gas plumes and its impact on the health of the population of Goma and its surroundings deserved special attention. He found that 75% of respondents said that volcanic plumes contain gases such as fluorine, ... 19.79% said that the volcanic plumes contained dust that was harmful to health, in particular slag and the hair of peelers.

He concluded by making a number of suggestions to the authorities and the general public: That the authorities intervene concretely by funding research so that they can intensify the monitoring of the volcano and Lake Kivu;

The authorities should encourage young university students to go abroad to study volcanology in greater depth, especially by granting them study bursaries; researchers should work day and night to predict all possible volcanic risks; the public should heed the advice of volcanologists in order to reduce the risks associated with the volcano.

- **Sarah SCAGLIONE at al (2011), the** aim of their studies was to measure the impact of volcanic deposits in the vicinity of the crater of the Nyiragongo volcano.

They found that the large quantity of volcanic gases and particles emitted (volcanic ash, peeling hair, slag) includes sulphur, halogens and trace elements, which strongly affect the chemistry of rainwater and have a large impact on the vegetation surrounding the volcano.

After analysing samples of rainwater collected from the edge of the crater, leaves of amanrantus viridis (amaranth), which is one of the main vegetables consumed by the local population, bags of moss were analysed. The sample of

rainwater collected at the edge of the crater had low PH values (≈3), high concentrations of F- and Cl- (up to 12.0 and 12.8mg/L, respectively) and dissolved toxic elements (talc Al, As, Cd, Cu, Fe and Pb). Biomonitoring results highlight that bioaccumulation of trace elements is extremely high near the crater rim and decreases with distance from active craters.
They conclude by saying that the wind, which blows regularly from east to west, carries the plume of ash and other materials in that direction. The ash destroys vegetation and can make the animals that graze on the grass along this route sick or even kill them, as these very harmful rock fragments can injure the stomach. To deal with these risks, we need to: wash vegetables before eating, cover food well during the season, keep livestock away from areas where ash falls, be careful with rainwater and only drink water treated by REGIDESO.

- **MAPENZI BALUME (2015), the** aim of his work was to raise awareness among the population of Goma of the risks associated with volcanological activities on investments in order to safeguard life and the socio-economic balance of the city of Goma.

The author found that the attitudes of the population of the town of Goma are linked to religiosity and myths about the volcano, such as: the volcano can't erupt where it used to. It's a matter for the ancestors who decide whether the volcano will erupt or not. Volcanoes don't erupt everywhere at once.
Following the survey, the author came to the conclusion that 64% of the population believe that raising awareness and providing information about volcanic risks in the city of Goma could prevent and/or reduce them. By adopting appropriate attitudes and putting in place management tools such as a contingency plan in the event of a crisis. The author concludes by proposing a project to raise the awareness of the population of the town of Goma about volcanic risks, at an overall cost of $890,805,85 per year, in order to improve practices with regard to volcanic risks. He therefore calls on everyone - the government, the OVG, the local population and the donors - to provide technical, financial and material support for the project.

- **Henry GAUDRU (2008)**, the aim of this study was to assess the risks to people living near the volcano.

He noted that around 50 to 60 volcanoes erupt every year around the world. The largest eruptions are life-threatening and destroy populated areas. Most of the most dangerous volcanoes are in Third World countries, as the number of victims shows.
It found that only pyroclastic flows accounted for 32% of victims worldwide, while volcanic ash, lahars, tsunamis, volcanic gases, famine and unknown hazards accounted for 21%, 17%, 19%, 6%, 1% and 4% respectively of victims

worldwide.
He concludes by saying that mitigating volcanic risks is possible, but it requires a combination of many elements, in particular: scientific prevention, raising public awareness, and he goes on to say that, while studying, forecasting and warning is the primary task of scientists, we must also make an effort to educate officials, the media and the local population about the appropriate measures. All these factors will help us to significantly mitigate the effects of volcanic eruptions, and save human lives too.

- **NDULU JUAKALI Mediator (2016);** her objective was to contribute to improving the roles of the Goma Volcanological Observatory (OVG).

He noted that the dangers of the volcano are to be feared before, during and after the eruption, because volcanic lava spares no one as it passes, not forgetting the effects of the various emanations such as gases and volcanic ash, which is why improving the role of the Goma volcanological observatory is of vital importance.
He found that 41.6% of respondents said that the role of the OVG was to evacuate the population to suitable areas and 31.8% felt that the OVG's role was to raise awareness of volcanic risks.
He concludes by outlining strategies to ensure that the OVG is equipped with surveillance equipment that is secure to prevent vandalism, and that staff are properly motivated in relation to their salaries.

- **Julien LUKUBIKA (2019);** the aim of his work was to determine the knowledge, attitudes and practices of households in the city of Goma in the Majengo district on the risks associated with the proximity of the Nyiragongo volcano. The author found that, of the seven (7) risks associated with the proximity of the volcano, including lava flow, volcanic ash and volcanic projection, glowing cloud, volcanic gases, mud flow (lahar), instability and tsunami. The only hazards known to most households are the glowing cloud, volcanic projection and tsunami, hence the low level of knowledge among households in the Majengo district about the risks associated with the proximity of the Nyiragongo volcano.

Following the survey, the author found that 33.4% of households were completely dissatisfied with what they were doing to limit the damage caused by the risks in the future. Hence the need to develop a relationship with scientists and the political and administrative authorities, to raise awareness and mitigate the risks associated with the proximity of the Nyiragongo volcano.
From these documents, we can see that we are not the first researchers to have carried out research on volcanism, but our research topic differs from those of others in that it involves studying the **knowledge, attitudes and practices of**

the inhabitants of Rusayo with regard to the health effects of the volcanic ash emitted by the Nyiragongo volcano.

0.1. Issues

Around 600 million people worldwide live in areas potentially affected by volcanic hazards. During a volcanic crisis, people may be evacuated to protect them from life-threatening hazards (e.g. pyroclastic flows), but they may still be exposed to potentially dangerous atmospheric volcanic emissions. Volcanic ash is a ubiquitous hazard, potentially distributed over thousands of square kilometres. Inhaling ash can exacerbate existing asthma and bronchitis symptoms, as well as respiratory symptoms such as coughing and shortness of breath. However, the danger to respiratory health from volcanic ash depends on its physicochemical composition, which can vary considerably from one eruption to another. **(Judith Coveya at al., 2019)**

Because of the risks to airliners flying through a volcanic ash cloud, the International Civil Aviation Organization introduced a Volcanic Watch for international air routes in 2002. The fine ash particles ejected at high altitude during eruptions can damage aircraft engines and electronics and reduce pilots' visibility (**Canthy Clerbaux at al.** ***2011).***

Residents of the Yogyakarta Special Region endured the events of 2010; 'Centenary' eruption of the Merapi volcano (located just 30km north of Yogyakarta), during which substantial ash fell on communities, on this occasion PMI Yogyakarta (local branch of the Indonesian Red Cross) distributed around 1 million basic face masks. **(Damby at al., 2013)**

During periods of volcanic activity, communities in Kagoshima City and the surrounding rural districts of Kihoku, Ushine, Kaigata and Sakurajima Island itself have been exposed to frequent ash falls. The city has well-organised ash removal practices, including the regular use of ash sweepers and the distribution of yellow plastic bags to people to collect ash from their properties. Although the local government of Kagoshima City recommends a reduction in exposure, including refraining from going outside and wearing a mask in the event of heavy ash fall. Although the local government is stockpiling masks for use in the event of a flu pandemic, masks are not being systematically distributed. Consequently, if residents or visitors wish to use them, they must obtain their own. **(Judith Coveya at al., 2019)**

Each volcano is unique and the quantities of sulphur dioxide and volcanic ash emitted vary, as do the heights at which they are injected into the atmosphere. SO2 is often emitted before an eruption and can therefore be used as an indicator to warn of an imminent eruption. Eruptions can be accompanied by volcanic ash. Every year, several eruptions are detected by IASI (*Infrared Atmospheric*

Sounding Interferoleter). Their impact on aviation varies according to the geographical area affected, the altitude at which the SO_2 was injected, the presence of ash and the persistence of the event. The months of May and June 2011 were marked by a succession of three volcanic eruptions, putting the centres responsible for airspace surveillance on constant alert. **(Canthy Clerbaux at al.** ***2011)***

The cloud of volcanic ash caused by the eruption of a volcano in Iceland poses a real threat to aircraft passing through.

Several thousand passengers were stranded on the ground after a large part of northern European airspace was closed due to a huge cloud of volcanic ash caused by a violent volcanic eruption in Iceland. **(Teade NEIL 2010)**

Ash from a volcano atop the Eyjafjallajokull glacier, the second major volcanic eruption in less than a month, has spread eastwards across the Atlantic, forcing the closure of airspace more than 1,700 kilometres away. The United Kingdom, Denmark, Norway and Sweden closed their airspaces. For their part, Belgium, France, Finland, Germany, the Netherlands and Spain suffered numerous disruptions to their air traffic. Volcanic ash poses a real threat to aircraft engines," explains geologist Teade Neil, "and any plane flying from Europe to the United States could take the risk of flying through a cloud of volcanic ash, and that's not a risk worth taking" (**Idem**).

Ash fall was substantial in the small farming community in southern Iceland near the volcano (up to 4 cm was deposited in the lowlands south of the volcano on 17 April). During the eruption, there was little precipitation in the region and ash deposits continued to be re-suspended by wind and human activity for months after the eruption had ended. Ambient air monitoring stations have been set up by Icelandic scientists in the affected areas. **(C. J Horwel at al. ; 2013)**

Volcanic ash tends to settle in the atmosphere at an altitude of 11,000 metres, the normal cruising altitude of an aircraft. The ash is so dangerous because of the fine dust that is almost invisible to the naked eye. This is why pilots fly through the ash cloud without realising it, turning on the suction of nanoparticles of volcanic dust in aircraft engines, which will clog the engines.

Back in 1982, when a British Airways flight from Kuala Lumpur (Malaysia) narrowly avoided disaster after passing through a volcanic ash cloud over the Pacific, the incident prompted the *aviation industry* to rethink the way it deals with ash clouds. It is thanks to this that international contingency plans have been activated as soon as an ash cloud is identified by a satellite, and airspace is closed as a preventive measure. **(Teade NEIL; 2010)**

According to Teade Neil (2010), it is difficult to predict where and for how long airspace will be disrupted by volcanic ash. This will depend on how violent the volcanic eruption is, how long the volcano spews ash into the air, and how long the wind continues to blow the ash cloud.

The eruption of the Nabro volcano on 12 June 2010, on the border between Ethiopia and Eritrea, has had adverse effects on the Ethiopian population. The volcano erupted on 12 June 2010, spewing volcanic ash over several hundred kilometres, contaminating water sources and disrupting air traffic in certain regions. As the Ethiopian population did not have the appropriate attitudes and practices to deal with the ash fallout, the negative repercussions of the eruption led to an increase in the number of reported cases of livestock mortality, displacements, health problems among the population, as well as a serious water shortage and a resurgence of malnutrition in the Woredas (administrative division) most affected by the eruption, according to a report by the Office for the Coordination of Disaster Prevention and Food Security Programmes in Afar, Ethiopia. **(Annal of Ethiopia 2010/Vol 25)**

The NYIRAGONGO volcano (Democratic Republic of Congo) belongs to the Virunga volcanic chain and is one of the most active volcanoes in Africa. It alone produces between 7,000 and 50,000 tonnes of pollutants (gas plumes, volcanic ash, etc.) per day. This pollution record is estimated at almost the entire emissions of the European Union in one month **(Dario Tedesco at al.2016).**

According to a study by **the University of Florence, OVG and GEVA (2005)**, the activity of the Nyiragongo volcano has a serious impact not only on the flora of the summit but also on that of all the surrounding regions, especially those located to the north-west, south-west and south of the volcano. Both wild and domestic plant species (cultivated by farmers) are attacked by volcanic deposits, and rainwater is contaminated by them. In this study, they also state that cultivated plants are more vulnerable than wild plants; this causes a serious problem for the food resources of the population in the areas concerned, but also for the entire economy of the villages surrounding the volcanoes, since cultivated plants are not only consumed as vegetables by the local population but are also sold in urban markets.

Around 40,000 to 50,000 people living around the Nyiragongo volcano depend on rainwater collected in cisterns. **(Jacques Durieux at al.; 2011**) The same water that burns vegetation is what people drink. In addition, high levels of fluoride emanating from the two volcanoes were also found in the drinking water.

In RUSAYO in the DRC, a village located to the south-west of the Nyiragongo volcano, the inhabitants do not have sufficient knowledge about the volcanic ash plume (practices and attitudes to adopt when volcanic ash falls) and volcanic ash is one of the most harmful effects on people living near active volcanoes, as it contains several chemical elements that are harmful to human health, such as fluorine, sulphur dioxide, etc. The RUSAYO group is one of the poorest villages, with the inhabitants farming as a source of income and basic food (vegetables, for example), while the vegetation in this area is more often than not invaded. The RUSAYO group is one of the poorest villages, and the inhabitants rely on agriculture as a source of income and staple food (vegetables, for example), while the vegetation in this area is more often than not invaded by volcanic deposits and, more often than not, volcanic ash and peeled hair, causing a number of illnesses among the inhabitants. As rainwater is also contaminated by volcanic emissions, the inhabitants of RUSAYO still rely on rainwater as their main source of drinking water, unaware that rainwater contains several contaminants from volcanic emissions, including volcanic ash suspended in the atmosphere, which in turn contains a high concentration of fluoride. By drinking rainwater, the residents of RUSAYO are exposing themselves to high levels of fluoride, which causes fluorosis. This fluorosis attacks not only the teeth but every bone in the skeleton. An increased number of deaths in cattle and goats is suspected to have been caused by the consumption of ash-coated grass, leading to deterioration of the animals' digestive systems and intoxication by fluorosis or excess fluoride.

When asked about the effects of volcanic ash on human health in his health area, the head of the Rusayo health centre, **Dr Jean Bosco**, said that local doctors had reported an increase in the prevalence of digestive problems and diarrhoea, a common effect of sulphur. Local doctors have also noted an increase in respiratory ailments among their patients during periods when the Nyiragongo volcano releases ash into the atmosphere. Farmers also claim that their harvests fail during the period when the Nyiragongo volcano releases a quantity of volcanic ash into the atmosphere; the volcanic ash, by settling on the various crops, particularly beans, potatoes, manioc and others, has a negative impact on their harvests. The plants most affected are those most essential to food security, such as beans, bananas, potatoes, sweet potatoes and maize.

In a press release, the Observatoire Volcanologique de Goma (OVG) has reassured the population of Goma and the surrounding area that the volcanic ash that fell over a large part of the area around the Nyiragongo volcano on the afternoon of Saturday 24 July 2021 is the result of the collapse of part of the crust of the central crater of the Nyiragongo volcano. It also calls for people to observe the rules of hygiene by carefully covering food and washing up, and to refrain from using rainwater as drinking water.

(www.grandlacksnews.comm/l-ovg-rassure-après-la-chute-des-cendres-volcanoes-in-the-zone-around-the-volcano-nyiragongo)

When ash falls, communities need rapid and accurate information on likely respiratory impacts from reliable sources, such as government agencies (e.g. civil protection/emergency management or health), but this is still hampered by the lack of immediate scientific information on the likely risk of the ash. **(C.J. Horwell at al. 2017)**

In view of the above, we asked ourselves a series of questions:

❖ **Main question**:

How much do the people of RUSAYO know about the health effects of the volcanic ash emitted by the Nyiragongo volcano?

❖ **Specific questions**

1. What are the attitudes of the people of RUSAYO to the volcanic ash emitted by the Nyiragongo volcano?
2. How do the people of RUSAYO protect themselves from the volcanic ash emitted by the Nyiragongo volcano?
3. What is the solution for mitigating the health effects of the volcanic ash emitted by the Nyiragongo volcano?

0.2. Assumptions of the work

❖ **Main hypothesis:**

The people of RUSAYO are said to have little knowledge of the health effects of the volcanic ash emitted by the Nyiragongo volcano;

❖ **Specific assumptions:**

1. The attitude of the people of RUSAYO to the volcanic ash was to panic and leave the fallout zone, taking it as a normal phenomenon;
2. To protect themselves from the volcanic ash, the people of RUSAYO keep their livestock away from the fallout zone, drink rainwater and do nothing special in the event of volcanic ash falling;
3. The construction of standpipes in the RUSAYO group and/or capacity building for the people of RUSAYO in terms of volcanic ash would be a solution to this problem.

0.3. Objectives of the work

❖ **Overall objective**

Reducing the risk of health effects from the volcanic ash emitted by the Nyiragongo volcano in the RUSAYO group.

❖ **Specific objectives**

1. To assess the attitudes of the inhabitants of RUSAYO to the volcanic ash emitted by the Nyiragongo volcano;
2. To determine the practices of the inhabitants of RUSAYO with regard to the

volcanic ash emitted by the Nyiragongo volcano;

3. Develop a project to build standpipes in the RUSAYO group and/or build the capacity of the people of RUSAYO to deal with the volcanic ash emitted by the Nyiragongo volcano.

0.4. Choice and interest of subject

a. Choice of subject

The choice of our subject was motivated by observations made in the field during research at the Goma Volcanological Observatory (OVG). When we saw the socio-sanitary situation of the inhabitants of Rusayo (one of the villages located near the Nyiragongo volcano) during our various field trips, these inhabitants being exposed for long periods to the plumes of volcanic ash and not knowing what to do before, during and after the fall of volcanic ash, we felt it necessary to guide the population through this study.

b. Interest in the subject

- **Scientific:** the results of this work will inform and form a database for future researchers in this field;
- **Social:** to encourage the population to become more aware of the damage caused by the volcanic ash emitted by the volcano, and to enable those working to help the population living near the volcano to carry out their activities in a way that reduces the risks to the population from living next to an active volcano;
- **Personal:** This work is of great interest to us because it will help us to deepen our knowledge of community problem management by providing effective and participatory solutions.

0.5. Spatial, temporal and material delimitation

We have delimited our study in time and space in order to avoid an all-encompassing character that could ultimately give us biased results.

The spatial axis includes the RUSAYO Group in NYIRAGONGO territory in North Kivu DRC. The timeframe is from January to November 2022.

0.6. Subdivision of work

In addition to the Introduction and General Conclusion, this work has three(3) chapters, including:

Chapitre 1. PRESENTATION OF THE RUSAYO GROUP AND GENERAL INFORMATION ON THE KNOWLEDGE, ATTITUDES AND PRACTICES OF THE INHABITANTS OF RUSAYO WITH REGARD TO VOLCANIC ASH

Chapitre 2. METHODOLOGICAL APPROACH, PRESENTATION AND DISCUSSION OF SURVEY RESULTS

Chapitre 3. DEVELOPMENT PROJECT FOR THE CONSTRUCTION OF

STANDPIPES IN THE RUSAYO DISTRICT

0.7. Research constraints

Given that there is no rose without thorns, carrying out this work was not as easy a task as one might think. The lack of specific libraries in the area, difficult access to certain data, and poor Internet connections made research difficult. These were the difficulties we had to face throughout our study.

PARTIAL CONCLUSION

The aim of this chapter was to look back at other researchers who have developed a theme similar to ours in order to gain a better understanding of the level of knowledge, attitudes and practices of the inhabitants of RUSAYO with regard to volcanic ash. We cited the researchers we had consulted and then presented our problem, which enabled us to present the dangers faced by the inhabitants of RUSAYO, which in turn led us to ask questions to which we proposed provisional solutions. In this same chapter, we showed the objectives and interests of our subject.

CHAPTER I

PRESENTATION OF THE RUSAYO GROUP AND GENERAL INFORMATION ON THE KNOWLEDGE, ATTITUDES AND PRACTICES OF THE INHABITANTS OF RUSAYO WITH REGARD TO THE VOLCANIC ASH EMITTED BY THE NYIRAGONGO VOLCANO

INTRODUCTION

This section looks at some important information about the RUSAYO group, as well as a review of the literature and key concepts on the subject (knowledge, attitudes and practices of the inhabitants of Rusayo regarding the health effects of the volcanic ash emitted by the Nyiragongo volcano).

I.1. DESCRIPTION OF THE RUSAYO GROUP

1.1.Geographical location

The RUSAYO group is located in the Nyiragongo territory, in the Bukumu chiefdom of South Kivu in the Democratic Republic of Congo.

It is limited:

- ✓ In the North: Virunga National Park;
- ✓ To the south: the town of Goma /commune of Karisimbi ;
- ✓ To the east by: Groupement Mudja; and
- ✓ In Oust by: the Kamuronza Group.

(Source: Administrative Secretary of the Rusayo group)

a. Soil and vegetation

The Rusayo group has a volcanic soil, derived from the weathering of volcanic lava and ash; a soil with good fertility.

The dominant vegetation in the Rusayo group is eucalyptus.

1.2.Political and administrative aspects

The Rusayo group is headed by a group leader, Mwami **Janvier KAMBUMBA BANGUMYA.**

The Rusayo group has 7 villages and each village has a village chief and a notable, including :

N°	Village	Village chief
O1	KARAMBI	MAPINDUZI MUREFU
02	RUKORWE	NDYANABO MUNIHIRE
03	KALANGALA	MOISE KANANE
04	KATWA	SAFARI MAHANGA
05	SHANGUTA	NTAWIHEBA MBANZA
06	KAHANDE	BUZIMANA BANGUMYA
07	KABALE-KATAMBI	KABUMBA MUREFU

Since the creation of the Rusayo group in **1945,** Rusayo has had 4 group leaders:

N°	Name and post name	Period
01	KAHANDE KAMIRIMO	1945-1960
02	BANGUMYA BAHIYI	1960-1994
03	Jean-Pierre BANGUMYA	1994-2009
04	Janvier KAMBUMBA BANGUMYA	2009 to present

1.3.Demographics

The Rusayo grouping is home to a number of tribes:

✓ The Bakumu ;
✓ The Nande ;
✓ The Batembo ;
✓ The Hunde ;
✓ Etc.

The Rusayo district has a total **population of 28412.**

Men Women Youth Girls Total Boys	
Population local	**80439378493960 5228412**

(Source: Rusayo group census taker; July 2022)

I.2 GENERAL INFORMATION ON THE KNOWLEDGE, ATTITUDES AND PRACTICES OF THE INHABITANTS OF RUSAYO REGARDING THE VOLCANIC ASH EMITTED BY THE NYIRAGONGO VOLCANO

1.2.1. Definition of key concepts

a) knowledge: faculty of knowing; way of understanding, of perceiving. (*LAROUSSE: encyclopaedic dictionary)*

In philosophy, knowledge is the state of knowing or knowing something. Knowledge" also refers to things that are known themselves, and by extension to things that are considered to be knowledge by a given individual or society. (https://fr.wikipedia.org/wiki/Connaissance philosophy)

b) attitude: way of behaving (*LAROUSSE: dictionary).*
encyclopaedic)

For **Thomas and Znaniecki**, an attitude is always object-oriented. It makes it possible to predict an individual's actual and potential behaviour in the face of social stimulation. (https://www.universalis.fr-encyclopedie-attitude-1 -le-conceptd-attitude)

c) practice: the act of carrying out a concrete activity; experience, in-depth habit. (*LAROUSSE: encyclopaedic dictionary*)

Application, execution, putting into action the rules and principles of a science, technique, art, knowledge, etc., as opposed to theory. (https://www.larousse.fr/dictionnaires/francais/pratique/63257)

d) Inhabitant: a person who ordinarily lives in a place; a human being or animal who settles in a place (*LAROUSSE: encyclopaedic dictionary*).

e) Volcanic ash: Volcanic ash is rock and mineral fragments less than 2 mm in diameter, ejected by a volcano. These particles are so fine that they can travel hundreds of kilometres and fall back to earth in the form of ash showers. (https://fr.wikipedia.org/wiki/Cendre volcano/definition)

f) volcano: From the Latin vulcanus, vulcaine=god of fire; conical mountain resulting from the accumulation of matter from the bowels of the earth, which rises through a crack in the earth's crust (chimney) and exits through a circular opening (crater). **(LAROUSSE: Maxipoche 2009)**

The place where products (gases, liquids and solids) of deep magma origin, which may be terrestrial or submarine, are released onto the surface (www. georisques.gouv.fr/dossier/volcanism).

g) volcanism: All volcanic activity. **(LAROUSSE: Maxipoche 2009)**

h) volcanic: Relating to volcanoes **(LAROUSSE : Maxipoche 2009)**

1.2.2. Theoretical approach

1. 2.2.1. KNOWLEDGE OF VOLCANIC ASH

1.1.Composition of volcanic ash

Unlike combustion ash, volcanic ash is hard and abrasive. It does not dissolve in water and conducts electricity well, especially when wet. During an ash shower, the sky appears hazy or yellowish and a sulphurous odour hangs in the air. The composition of volcanic ash clouds varies from one volcano to another. Generally speaking, however, it is composed mainly of silica (> 50 per cent), with smaller quantities of aluminium, iron, calcium and sodium oxides. Silica comes in the form of vitreous silicates and, under a scanning electron microscope, looks like sharp-edged shards of glass. The vitreous silicate material is very hard, generally hardness level 5 or 6 on the Mohs1 scale (similar to a typical penknife blade), with a proportion of material hardness equivalent to quartz (level 7), all in powdered form is extremely abrasive. **(International Civil Aviation Organization, 2007)**

Constituent	Fuego, 1974	Mount St. Helens Helens, 1980	El Chichon, 1982	Galunggung, 1982
		Percentage by weight		
SiO2	53.30	71.40	68.00	61.30
Al2O3	18h70	14h60	15.90	7.10
Fe2O3, FeO	9.10	2.40	1.60	7.10
CaO	9h40	2.60	2.12	5.70
Na2O	3.90	4h30	4.56	4.00
MgO	3.40	0.53	0.25	4.00
K20	0.80	2.00	5.05	1.70

TiO2	1.20	0.37	0.29	1.3
P205	-	0.99	0.00	E0.33

Figure 1: Table showing the Composition of ash particles found in ash clouds from eruptions of four volcanoes (International Civil Aviation Organization, 2017).

1.2.Dangers of volcanic ash

Breathing volcanic ash can cause problems for people with respiratory problems. Their abrasive surfaces can cause irritation to the skin and mucous membranes. The combination of ash and moisture in the lungs can turn it into a liquid cement that can make breathing difficult. For this reason, it is advisable to breathe through a cloth or mask (https://fr.wikipedia.org/wiki/Cendre volcanic).

a) Health effects of volcanic ash

The effects of ash on health can be divided into several categories: respiratory effects, eye symptoms, skin irritation and indirect effects.

1. Respiratory effects

In some cases, eruptions can send very fine ash into the lungs. At high levels of exposure, even very healthy people can become ill with the following acute symptoms:

o Nasal irritation with runny nose. Throat irritation and angina, sometimes accompanied by a dry cough.

o People with chest pain could develop severe symptoms of bronchitis lasting a few days beyond exposure to the ash (production of sputum, wheezing, shallow breathing).

o Airway irritation for people with asthma or bronchitis.

o Breathing becomes uncomfortable.

Fine ash also causes the lining of airways to produce more secretions, which can make people cough and breathe more difficult. Asthmatics suffer, especially children who are heavily exposed to ash when they play. They may suffer from coughing, chest tightness and wheezing.

2. Eye symptoms

Eye irritation is a common health effect, as a piece of gravel or sand can cause painful scratches in front of the eye and conjunctivitis.

Common symptoms include:

o The eyes feel a foreign body inside them, become painful, disturbing and infected with blood.

o Show viscous tearing or ripping.

o Corneal abrasion or scratching

o Acute conjunctivitis or inflammation of the conjunctival sac surrounding the "eye ball" due to the presence of ash, leading to redness, burning of the eyes and photosensitivity.

3. Indirect health effects

The indirect health impacts of large-scale ash fallout must also be considered. For example:

Falling ash clearly reduces visibility and can cause road accidents. Wet or dry fine ash is slippery and reduces traction. Depending on the amount of ash that falls, it can make the road impassable and cut off the community from everyday life.

The ash plume can be dangerous when an aircraft passes close by. The fine ash enters the turbines, prevents the engines from operating and can cause a crash landing.

4. Special precautions for children

When the ashes fall, follow the instructions below for the safety of children:

- Keep children at home if possible.
- Advise children not to play outside or run around to avoid ingesting ash into their lungs.
- When children are outside during the ash fall, they must wear an approved mask. Otherwise, they should seek shelter until further instruction.

(International Volcanic Health Hazard Network (IVHHN) at Al: 2017)

b) Protecting yourself from volcanic ash

Protect yourself with a dust mask or improvised cloth. People with asthma, bronchitis or emphysema are advised to stay at home. Protect your eyes with goggles. Avoid going outside when you don't have to. Alternatively, use an umbrella. Limit driving and, if necessary, take an asthma medication.

Safety distance.

Immediately after ash fallout, visibility and air quality may be affected by the suspension of fine ash. Falling ash can make water undrinkable. You should therefore stock up on water before the eruption. If tap water is used, do not drink it during and after the eruption unless contraindicated by the safety services. Cover open cisterns.

Lightly wet the ash before cleaning to avoid problems of overloading the roof and excessive exposure to dry ash.

Volcanic ash can contaminate water, particularly cistern water. Most cisterns are not covered and the quantity of ash, however small, can pose drinking water problems. During and after the fall of ash, cisterns must be cleaned if the ash contains high levels of fluoride (1.5mg/l, the threshold set by the WHO). Water containing this element, when drunk, can cause a disease known as "fluorosis".

The ash load can cause roof collapses, injuries and even death. During eruptions, people have lost their lives by falling from their roofs trying to clear the ash.

Animals are exposed to the danger of poisoning when they graze on grass covered with ash containing hydrofluoric acid **(Judith Covey at al.2019).**

c) Action to be taken in the event of volcanic ash fallout

- o Close windows and doors if ash falls.
- o Plug any cracks that could allow ash to escape. Protect electronic equipment until all the ash has been removed.
- o Prevent ash from entering household ducts, but allow ash water to drain away.
- o Cover cisterns; stock up on water before the ash falls and do not use tap water during the fall, especially if it is raining.
- o People with bronchitis, emphysema or asthma should remain sheltered and avoid exposure to ash. Make sure there is water and healthy food in stockpiles.
- o Know your child's school emergency plan.

d) What to do in the event of ash fallout

- o Don't panic, stay calm.
- o Stay at home or in the office - take shelter. Use a mask, handkerchief or cloth to cover your nose and mouth.
- o If the alert is given before the fall begins, return home.
- o Do not use the telephone if there is no emergency.
- o Listen to the radio to follow the progress of the rash. Do not wear contact lenses to avoid abrasion of the cornea.
- o If there is ash in the water, let it settle and use only clear water. If there is a lot of ash in the water, do not use your washing machine. Water contaminated by ash (the water is cloudy) may present a health risk.
- o Eat vegetables and rhizomes once they have been thoroughly washed.

e) Cleaning volcanic ash

Volcanic ash is different from normal dust. It has a crystalline structure and is therefore abrasive on any surface to which it is wiped. Volcanic ash is a problem because it can be found everywhere in the home, office, electronic and computer equipment and pipes, causing irreparable damage. The ash is remobilised by wind and rain, which can transport it to previously cleaned or unaffected areas.

Caution! Take precautions when cleaning up the ashes. Wear masks and goggles. Wet the ash lightly. Do not let the ash accumulate over several centimetres on the roof of your house, as there is a risk of it collapsing. When using a ladder, be very careful as the ash makes the surfaces slippery.

- o Close off all houses, cars, machinery and pipe systems and clean up the ash at the same time. Use a shovel and a stiff broom. Put the ash in a plastic bag or truck.

o Use less water to clean the site. However, the arrival of rain will prevent the ash from rising but will drain a massive quantity into the pipe system and out to sea.

o Shake out clothes before entering the house or office. Do not dump ash in the garden or by the roadside. Do not clean ash into gutters or drains. This can damage the water treatment system and block the pipes.

o Use an appropriate hoover for carpets, furniture, office equipment, household appliances and other items, or a compressed air gun depending on the material to be cleaned.

o Don't drive unless you have to.

o If there is an emergency, gently roll over and clean the windscreen wipers with bottled water and a cloth.

o Change the engine oil, oil and air filters every 80160km if the ash is too thick. Wash the car from the inside of the engine to the seats and tyres.

(**International Volcanic Health Hazard Network (IVHHN) at Al. 2003**)

1.3.Precautions to be taken by residents living near the volcano to protect themselves from volcanic ash

When ash falls, communities need rapid and accurate information on the likely respiratory impacts from reliable sources, such as government agencies (e.g. civil protection/emergency management or health), but this is always hampered by the lack of immediate scientific information on the likely risk of the ash. Each volcanic explosion is different, in terms of the health-relevant characteristics of the ash generated (even from the same volcano and within the same eruption sequence). While the International Volcanic Health Hazard Network (IVHHN, www.ivhhn.org) has developed protocols for the rapid characterisation of ash for health risk assessment. The reality is that this detail cannot generally be provided while the ash is still falling **(Damby et al., 2013).**
Given that such analyses will only provide an indication of the potential respiratory risk and that epidemiological/clinical studies can take months or even years, the World Health Organization/Pan American Health Organization is taking a precautionary approach and has developed generic advice to be used globally by health/disaster management agencies and NGOs on protecting the community when ash is airborne. *They recommend staying indoors, but if you must be outdoors, use a mask, handkerchief or "simple" cloth.*

A handful of studies have examined the effectiveness of respiratory protection for general public use (i.e. not related to volcanic eruptions) and have tested the filtration performance of masks or common fabric materials against ultrafine particles (smaller than 1m; equivalent to certain influenza pathogens, which are much smaller than most ash particles) or fine-grained particles (1.0 to 2.5m;

which simulate larger pathogenic particles such as viruses, urban pollutants, allergens and construction dust). **(Rengasami at al.; 2004**) tested the filtration efficiency, against NaCl aerosol (0.02-1.0m), of five fabrics **(C.J. Horwell at al. 2017)**

I.2.2.2. Plume direction of the volcanic ash from Nyiragongo and Nyamulagira

Figure 2: Aerial view of the plumes of ash and volcanic gases escaping from the crater of the Nyiragongo and Nyamulagira volcanoes on 9 February 2015, captured by the Goma Volcanic Observatory (Charles M. Balagizi at al. 2016).

I.2.2.3. Volcanic ash and its effect on water and health in the vicinity of the Nyiragongo and Nyamulagira volcanoes

During the 2010 eruption, concerns were expressed by local residents about water quality and sensations of physical discomfort (e.g. nausea, bloating, indigestion, etc.) after drinking rainwater collected after the eruption began. We present the elemental and ionic chemistry of drinking water samples collected in the region on the third day of the eruption (5 January 2010). We identify a significant impact on water quality associated with the eruption, including a lower pH (i.e. acidification) and an increase in acid halogens (e.g. F(-) and Cl(-)), major ions (e.g. SO(4)(2-), NH(4)(+), Na(+), Ca(2+)), potentially toxic metals (e.g. Al(3+), Mn(2+), Cd(2+), Pb(2+), Hf (4+)), and particulate charge. In many cases, the composition of the water far exceeds World Health

Organisation (WHO) drinking water standards. The degree of pollution depends on (1) the direction of the ash plume and (2) the density of the ash plume. The potential negative impacts on health depend on the pH of the water, which regulates the elements and their chemical form that are released into the drinking water. **(Emilio C. at.al ;2012)**

The gas plumes from Nyiragongo and Nyamulagira are rapidly converted into acidic compounds, which lead locally to acid rain and air pollution, posing a potential danger to people, animals, agriculture and seriously damaging the environment. **(Cuoco et al. 2012)**

Constant access to clean water remains a major challenge for the DRC, particularly in and around the city of Goma, where volcanism is considerably deteriorating water quality. Rain dissolves the gases and ashes contained in the plumes and becomes highly acidic, with a pH as low as 2 on the crater and 4-5 in most villages. Certain elements are present in rainwater, surface water and groundwater at concentrations exceeding WHO guidelines and recommendations for drinking water (for example, fluoride is regularly present at values > 1.5 mg/L). **(Balagizi et al., 2018)**

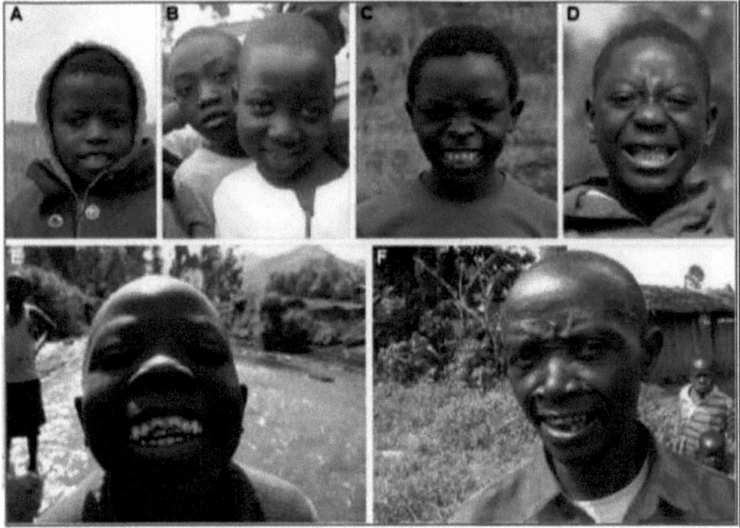

Figure 3: Photos showing young people and adults suffering from dental fluorosis in the villages around the Nyiragongo volcano, where rivers and rainwater contain fluoride in concentrations exceeding WHO guidelines and recommendations (Charles M. Balagizi at al. 2018).

PARTIAL CONCLUSION

In this chapter, we present the Rusayo group, define the key words of our research topic to help us understand our problem and review the literature;

CHAPTER II

METHODOLOGICAL APPROACH, PRESENTATION AND DISCUSSION OF SURVEY RESULTS

INTRODUCTION

This chapter will focus on the presentation of the data, analysis and interpretation of the survey results. It should be noted that the survey was carried out among the inhabitants of the RUSAYO group.

11.1. METHODOLOGICAL APPROACH

11.1.1. Research framework

This research is being carried out within a social, health and environmental framework, with the aim of ensuring the physical, mental and social well-being of the people of Rusayo through healthy, enriching and fulfilling conditions.

11.1.2. Type of search

In order to find out what the inhabitants of Rusayo know and do about the effects of volcanic ash on health, it is obvious to describe and evaluate the situation as experienced in the Rusayo Group. It is in this sense that this research is descriptive and evaluative.

11.1.3. Methods, techniques and tools

1. Methods

Method is a set of reasoned steps taken to achieve a goal **(Dictionnaire le Robert 2011)**. It is also defined as the set of intellectual operations used to analyse, understand and explain the reality being studied. **(Jean Louis LAUBET 2000)**

The methods used in this work are :

A. Descriptive method

The French language dictionary emphasises that description is the action of describing. This method was used generally in the description of the Rusayo group as a whole.

B. The analytical method

According to the Petit Robert, to analyse means to examine, to make an analysis. We used this method to analyse the knowledge, attitudes and practices of the people of Rusayo when faced with volcanic ash.

C. Statistical method

Statistics is defined as the science and technique of mathematical interpretation of complex and numerous data **(CT POLO FUETA Espérant 2022)**. This method enabled us not only to determine the sample for our study, but also to quantify the results of the research. It also enabled us to present the results in tables. By identifying the simple numbers and their respective percentages.

2. Techniques used

GRAWITZ. M (2000) defines a technique as a set of rigorous, well-defined procedures that can be applied at the same level and in the same conditions, depending on the type of problem. The techniques used in this work are :

A. Documentary technique

In preparing this work, we have drawn on a number of documents relevant to our subject, such as course notes, TFCs, dissertations, journals, books, archives and various reports and publications on certain addresses and sites.

B. Questionnaire technique

This technique enabled us to develop a written questionnaire that we administered to a sample representing our target community. It was very useful in testing our hypotheses.

11.1.4. Population and choice of sample

1. Study population

The study population for our work is the entire population of the Nyiragongo territory bordering the Nyiragongo volcano.

2. Target population

The target population for our work is all the inhabitants of the Rusayo group.

3. Sample size

In order to confirm our hypotheses and the objectives of this work, we used a representative sample, obtained by statistical calculation using the LUNCH formula, which is as follows:

$$n = \frac{NZ^2.p(1-p)}{Nd^2+Z^2\,p(1-p)}$$ Hence

n= Sample size

N= Study population equal to 28,412

Z= Coefficient of standard deviation of 1.96

p= the probability, which is 0.5

d= Margin of error either 10% or 0.1

So ;

$$n = \frac{28\,412.(1{,}96)^2.0{,}5(1-0{,}5)}{28\,412.(0{,}1)^2+(1{,}96)^2\,0{,}5(1-0{,}5)} = \frac{28\,412.3{,}84.0{,}25}{28\,412.0{,}01+3{,}84.0{,}25} = \frac{27275{,}52}{284{,}12+0{,}96}$$

$$=> \frac{27275{,}52}{285{,}08} = 95{,}67 \approx 96$$

n =96 Individuals distributed as follows:

60 individuals = local population ;

18 individuals= health workers from the Rusayo group and

18 individuals= OVG researchers

II.1.5. Data collection

Data collection took place in November 2022, using interviews and questionnaires. We carried out an initial field visit to observe the reality of our study. Following the actual survey, we administered a survey questionnaire to our respondents under our supervision, and for those who had difficulty reading or writing, the survey was conducted by interview.

II.2. PRESENTATION OF SURVEY RESULTS

II.2.1. Purpose of the survey

- Compare our hypotheses with the reality on the ground;
- Identify the health effects of the volcanic ash emitted by the Nyiragongo volcano in the Rusayo district;
- Find an effective, participative solution to the problem.

II.2.2 Interpretation of survey results

The interpretation of the survey results is presented in SPSS format:

1. IDENTIFICATION OF THE SURVEY

Table I: Sex of respondents

	Frequency	Valid percentage	Cumulative percentage
Valid Male	64	66,67	66,7
Female	32	33,33	100,0
Total	96	100,0	

Source: results of our 2022 surveys

The table shows that 64 of the 96 respondents (66.67%) were male, compared with 32 (33.33%) who were female.

Table II: Age of respondents

	Frequency	Valid percentage	Cumulative percentage
Valid for under 18s	20	20,83	20,8
18 to 25 years old	35	36,46	57,3
26 to 35 years	22	22,92	80,2
36 to 50 years	15	15,63	95,8
51 and over	4	4,17	100,0
Total	96	100,0	

Source: results of our 2022 surveys

The table shows that 20 of the 96 respondents, or 20.83%, were under 18; 35 respondents, or 36.46%, were aged between 12 and 25; 22 respondents, or 22.92%, were aged between 26 and 35; 15 respondents, or 15.63%, were aged between 36 and 50; and 4 respondents, or 4.2%, were aged 51 and over.

Table III: Civil status of respondents

	Frequency	Valid percentage	Cumulative percentage
Valid Married	60	62,50	62,5

Single	25	26,04	88,5
Divorced	7	7,29	95,8
Widow	4	4,17	100,0
Total	96	100,0	

Source: results of our 2022 surveys

The table shows that 60 of the 96 respondents (62.50%) were married, 25 (26.04%) were single, 7 (7.29%) were divorced and 4 (4.17%) were widowed.

Table IV: Position of respondents

	Frequency	Valid percentage	Cumulative percentage
ValidTeacher	15	15,63	15,6
Health agent	18	18,75	34,4
Cultivator	36	37,50	71,9
Student	9	9,38	81,3
OVG researcher	18	18,75	100,0
Total	96	100,0	

Source: results of our 2022 surveys

The table and graph show that 15 of the 96 respondents, or 15.63%, were teachers; 18 respondents, or 18.75%, were health workers; 36 respondents, or 37.50%, were farmers; and 18 respondents were OVG researchers.

2. RESULTS OBTAINED AMONG THE POPULATION LOCALE

A. Knowledge about the risks of volcanic ash

Table V: Have you ever heard of volcanic ash?

	Frequency	Valid percentage	Cumulative percentage
Valid YES	53	88,54	88,5
NO	7	11,46	100,0
Total	60	100,0	

Source: results of our 2022 surveys

The table shows that 53 out of 60 respondents (88.54%) had already heard of volcanic ash and 7 respondents (11.46%) had never heard of volcanic ash.

Table VI: If yes, what do you know about the volcanic ash emitted by the Nyiragongo volcano?

	Frequency	Valid percentage	Cumulative percentage
Valid Dust harmful to health	19	32,29	32,3
Affects rainwater	20	33,33	65,6
Affects vegetation	21	34,38	100,0
Total	60	100,0	

Source: results of our 2022 surveys

The table shows that 19 out of 60 respondents (32.29%) said that volcanic ash is a dust that is harmful to health, 20 respondents (33.33%) said that volcanic ash affects rainwater and 21 respondents (34.38%) said that volcanic ash affects vegetation.

Table VII: Do you observe volcanic ash in your environment?

	Frequency	Valid percentage	Cumulative percentage
ValidYES	60	100,0	100,0
Total	60		

Source: results of our 2022 surveys

The table shows that 100% of the 96 people surveyed said that they observed volcanic ash in the environment.

Table VIII: How do you know it's volcanic ash?

		Frequency	Valid percentage	Cumulative percentage
Valid	The smell	25	41,7	41,7
	Visible to the naked eye on vegetation	51	51,0	51,7
	From acid rain.	4	7,3	100,0
Total		60	100,0	

Source: results of our 2022 surveys

The table shows that 25 out of 60 respondents knew that volcanic ash was involved from the smell, 31 respondents (51.04%) knew that volcanic ash was involved from the visibility of ash on vegetation and 4 respondents (7.29%) said that they knew that volcanic ash was involved from acid rain.

Table IX: In your group, is volcanic ash visible permanently or periodically?

	Frequency	Valid percentage	Cumulative percentage
ValidPeriodically	60	100,0	100,0
Total	60		

Source: results of our 2022 surveys

The table shows that 60 respondents (100%) say that volcanic ash is periodically visible in their environment.

Table X: If periodically When?

	Frequency	Valid percentage	Cumulative percentage
Valid when there is activity in the central crater	39	65,63	65,6
Before the eruption	11	18,75	18,8

During the eruption	3	4,17	
After the eruption	7	11,46	100,0
Total	60	100,0	

Source: results of our 2022 surveys

The table shows that 39 respondents out of 60 (65.63%) said that volcanic ash was visible when there was activity in the central crater, 11 respondents (18.75%) said that volcanic ash was visible before the eruption, 3 respondents (4.17%) said that volcanic ash was visible during the eruption and 7 respondents (11.46%) said that volcanic ash was visible after the eruption.

Table XI: After the volcanic ash falls, do you know that you have to wash the vegetables before cooking?

	Frequency	Valid percentage	Cumulative percentage
Valid YES	60	100,0	100,0
Total	60	100,0	

Source: results of our 2022 surveys

The table shows that all 60 respondents, i.e. 100% of respondents, knew that they had to wash vegetables before cooking after the volcanic ash had fallen.

Table XII: After the volcanic ash falls, do you know that you should not drink rainwater?

	Frequency	Percentage	Valid percentage	Cumulative percentage
ValidYES	60	100,0	100,0	100,0
Total	60	100,0		

Source: results of our 2022 surveys

The table and graph show that 60 respondents, i.e. 100% of respondents, know that they should not drink rainwater if there is ash fallout.

Table XIII: What is the main source of water you use in Rusayo

	Frequency	Valid percentage	Cumulative percentage
Valid Collected rainwater in tanks	48	80,21	80,2
Lake water collected in cisterns by farmers	12	19,79	100,0
Total	60	100,0	

Source: results of our 2022 surveys

The table shows that 48 of the 60 respondents (80.21%) use rainwater collected in cisterns as their main source of water, and 12 respondents (19.79%) use lake water collected in cisterns by the farmers.

B. Attitude to volcanic ash fallout

Table XIV: What to do in the event of volcanic ash fallout

		Frequency	Valid percentage	Cumulative percentage
Valid	Panic and leave the area	11	17,71	17,7
	A normal phenomenon	7	11,46	29,2
	Remain calm and wait for instructions	42	70,83	100,0
Total		60	100,0	

Source: results of our 2022 surveys

The table shows that 11 respondents out of 60 (17.71%) said that they panicked and left the area when the Nyiragongo volcano released a quantity of volcanic ash into the atmosphere, 7 respondents (11.46%) said that this was a normal phenomenon and 42 respondents (70.71%) said that they remained calm and waited for instructions.

C. Inhabitants' practices regarding volcanic ash

Table XV: What to do in the event of volcanic ash fallout

	Frequency	Valid percentage	Cumulative percentage
Valid Staying at home and wait for the end of the ash fallout	1	1,04	1,0
Keep livestock away	28	46,88	47,9
from the drop zone Nothing special	31	52,08	100,0
Total	60	100,0	

Source: results of our 2022 surveys

The table shows that 1 respondent out of 60, i.e. 1.04%, stayed at home and waited for the ash to fall, 28 respondents, i.e. 46.88%, kept their livestock away from the fallout zone and 31 respondents, i.e. 52.08%, said they did nothing special in the event of volcanic ash falling.

D. What is the solution to mitigate the health effects of the volcanic ash emitted by the Nyiragongo volcano?

Table XVI: What is the solution to mitigate the effects of volcanic ash on health in Rusayo?

	Frequency	Valid percentage	Cumulative percentage
ValidateBuild standpipes in	53	88,54	88,5

the Groupement de Rusayo Capacity building for local residents of Rusayo in	7	11,46	100,0
volcanic ash Total	60	100,0	

Source: our 2022 field surveys

The table shows that 53 respondents out of 60 (88.54%) said that the construction of standpipes in the Rusayo Group would be a solution to the health effects of volcanic ash, and 7 respondents (11.46%) said that building the capacity of the people of Rusayo to deal with volcanic ash would be a solution to the health effects of volcanic ash.

3. RESULTS OBTAINED FROM HEALTH WORKERS

A. Knowledge about the risks of volcanic ash

Table XVII: Have you ever heard of volcanic ash?

	Frequency	Percentage	Valid percentage	Cumulative percentage
Valid YES	18	100,0	100,0	100,0
Total	18	100,0	100,0	

Source: results of our 2022 surveys

The table shows that 100% of the 18 respondents said they had heard of volcanic ash.

Table XVIII: If yes, what do you know about volcanic ash?

	Frequency	Valid percentage	Cumulative percentage
ValidDust harmful to health	13	72,22	72,2
Affects the water in the rain	5	27,78	100,0
Total	18	100,0	

Source: results of our 2022 surveys

The table shows that 13 out of 18 respondents (72.22%) say that volcanic ash is a dust that is harmful to health, and 5 respondents (27.72%) say that it affects rainwater.

Table XIX: After the volcanic ash falls, do you know that you should not drink rainwater?

	Frequency	Valid percentage	Cumulative percentage
Valid YES	18	100,0	100,0
Total	18	100,0	

Source: results of our 2022 surveys

The table shows that 100% of the 18 respondents knew that they should not drink rainwater after the volcanic ash had fallen.

Table XX: What is the main source of water you use in Rusayo

	Frequency	Valid percentage	Cumulative percentage
Validates Rainwater collected in tanks	6	33,33	33,3
Lake water collected in cisterns by farmers	12	66,67	100,0
Total	18	100,0	

Source: results of our 2022 surveys

The table shows that 12 respondents (66.67%) use lake water collected in cisterns by the farmers and 6 respondents (33.33%) use rainwater collected in cisterns.

B. Attitude to volcanic ash fallout

Table XXI: How would you react to volcanic ash fallout?

	Frequency	Valid percentage	Cumulative percentage
ValidatePanic and leave the zone	16	88,89	88,9
Remain calm and wait for instructions	2	11,11	100,0
Total	18	100,0	

Source: results of our 2022 surveys

The table shows that 16 respondents (88.89%) panicked and left the area and 2 respondents (11.11%) remained calm and listened to the instructions.

C. How local residents deal with volcanic ash

Table XXII: What do you do if volcanic ash falls?

	Frequency	Valid percentage	Cumulative percentage
Valid Cover nose and mouth once outside	16	88,89	88,9
Nothing special	2	11,11	100,0
Total	18	100,0	

Source: results of our 2022 surveys

The table shows that 16 respondents (88.89%) cover their nose and mouth when outdoors and 2 respondents (11.11%) do nothing special.

D. What is the solution to mitigate the health effects of volcanic ash?

Table XXIII: What is the solution for mitigating the effects of volcanic ash on health in Rusayo?

	Frequency	Valid percentage	Cumulative percentage
ValidatesBuilding terminals fountains in the Rusayo Group	18	100,0	100,0
Total	18	100,0	

Source: results of our 2022 surveys

The table and graph show that 18 respondents (100%) said that the construction of standpipes in the Rusayo groupement would solve this problem.

4. RESULTS OBTAINED FROM THE OVG

A. Knowledge about the risks of volcanic ash

Table XXIV: What do you know about volcanic ash?

	Frequency	Valid percentage	Cumulative percentage
Valid Dust harmful to health	10	55,56	55,6
Affects rainwater	7	38,89	94,4
Affects vegetation	1	5,56	100,0
Total	18	100,0	

Source: results of our 2022 surveys

The table and graph show that 10 out of 18 respondents (55.56%) said that volcanic ash is a dust that is harmful to health, 7 respondents (38.89%) said that volcanic ash affects rainwater and 1 respondent (5.56%) said that volcanic ash affects vegetation.

Table XXV: When is volcanic ash visible?

	Frequency	Valid percentage	Cumulative percentage
EnabledWhen there is activity in the central crater	15	83,33	83,3
After the eruption	3	16,67	100,0
Total	18	100,0	

The table and graph show that 15 out of 18 respondents (83.33%) said that volcanic ash was visible when there was activity in the central crater, and 3 respondents (16.67%) said that volcanic ash was visible after the eruption.

B. Attitude to volcanic ash fallout

Table XXVI: What do you do in the event of volcanic ash fallout?

	Frequency	Valid percentage	Cumulative percentage
Valid Stay calm and wait instructions	18	100,0	100,0
Total	18	100,0	

Source: results of our 2022 surveys

The table and graph show that 100% of the 18 respondents knew that they had to remain calm and wait for instructions.

C. Inhabitants' practices regarding volcanic ash

Table XXVII: What do you do in the event of volcanic ash fallout?

	Frequency	Valid percentage	Cumulative percentage
Valid Stay at home and wait for the end of the ash fallout	16	88,89	88,9
Nothing special	2	11,11	100,0
Total	18	100,0	

Source: results of our 2022 surveys

The table and graph show that 16 respondents out of 18 (88.89%) stayed at home and waited for the ash to fall, and 2 respondents out of 18 (11.11%) said they did nothing special when the volcanic ash fell.

D. What is the solution to mitigate the health effects of the volcanic ash emitted by the Nyiragongo volcano?

Table XXVIII: What is the solution for mitigating the effects of volcanic ash on health in Rusayo?

	Frequency	Valid percentage	Cumulative percentage
Valid Construction of terminals fountains in the Rusayo Group	6	33,33	33,3
Reinforcement of The capacity of the people of Rusayo to deal with volcanic ash	12	66,67	100,0
Total	18	100,0	

Source: our 2022 field surveys

The table and graph show that 6 respondents out of 18 (33.33%) said that the construction of standpipes in the Rusayo Group would be a solution to the health effects of volcanic ash and 12 respondents (66.67%) said that building the capacity of the people of Rusayo to deal with volcanic ash would be a solution to the health effects of volcanic ash.

II.3. DISCUSSION OF THE SURVEY RESULTS

With regard to the survey we carried out among the inhabitants of the Rusayo group in November 2022 to verify our hypothesis, what we found was that

From the results obtained from the local population, in **Table V** 88.54% have already heard of volcanic ash and in **Table VI** 34.38% say that volcanic ash affects vegetation and 33.33% say that it affects rainwater. In **Table X** 65.63% said that volcanic ash is visible when there is activity in the central crater. In the **Table XII** 100% of our respondents know that they should not drink rainwater in case of volcanic ash fallout. Whereas in **Table XIII** 80.21% use rainwater collected in cisterns as their main source. From this table we can directly understand that the inhabitants of Rusayo drink rainwater not because they want to but because they have no other source of drinking water accessible to all.

From all these tables we can see that the inhabitants of Rusayo have sufficient knowledge of volcanic ash, because even if we compare these results with those of **Bisimwa K (2013)** and **Sarah S at.al 2011,** we will see that our results are similar.

When asked about the attitudes of residents to volcanic ash fallout, 70.71% said they remained calm and waited for instructions, 17.71% panicked and left the area and 11.46% said this was a normal phenomenon; **Table XIV.**

So, despite the small number of respondents who said that volcanic ash fallout is a normal phenomenon, we can see that the people of Rusayo have a good attitude to volcanic ash.

Wanting to know the practices of the inhabitants of Rusayo regarding volcanic ash, only 1.04% stay at home and hear the end of the fallout of ash, 52.08% say they do nothing special in case of fallout of volcanic ash **Table XV.** Here we see that the inhabitants of Rusayo do not have good practices in the event of volcanic ash fallout.

Wanting to know the solution to mitigate the effects of volcanic ash on health; in **Table** XVI 88.54% said that the construction of standpipes in the Rusayo groupement would be the solution to this problem and 11.46 said that capacity building would be a solution to this problem.

From the results obtained from the health workers, in **Table XVII**, 100% of the health workers have already heard of volcanic ash. As for **Table XVIII**, 72.22% know that volcanic ash is a dust that is harmful to health and 27.72% say that it affects rainwater. In **Table XIX,** 100% of health workers know that they should not drink lake water collected in cisterns by farmers and 33.33% use rainwater collected in cisterns. From these tables we can see that the health workers have a good knowledge of volcanic ash.

In **Table XXI**, 88.89% of the health workers surveyed panicked and left the area, while 11.11% remained calm and listened to instructions. This table shows that health workers do not have the right attitude to volcanic ash fallout.

In **Table XXII**, 88.89% cover the nose and mouth once outside. This table shows that health workers have good practice in dealing with volcanic ash.

In **Table XXIII**, 100% of health workers say that the construction of standpipes would be the solution to this problem.

From the results obtained from the OVG researchers, in **Table XXIV** 55.56% of the OVG researchers surveyed say that volcanic ash is a dust that is harmful to health, 38.89% say that it affects vegetation.

In **Table XXV** 83.33% say that volcanic ash is visible when there is activity in the central crater and 16.67% say that it is visible after the volcanic eruption.

In **Table XXVI,** 100% of the researchers surveyed remained calm and waited for instructions. In **Table XXVII**, 88.89% stayed at home and heard the end of the ash fallout. From this we can understand that the OVG agents have good practice in dealing with volcanic ash.

In **Table XXVIII, 66.67%** said that building the capacity of the inhabitants of Rusayo to deal with volcanic ash was the only solution to this problem.

11.3. RECOMMENDATIONS

1. **To the State**: to take action to find funds to build standpipes to help these people so that they can no longer use rainwater as their main source of water.
2. **Health workers**: to continue to raise awareness among the people of Rusayo about the health effects of volcanic ash and to continue to show them that rainwater has a major impact on health because it contains several chemical compounds from the Nyiragongo volcano.
3. **To the local population**: to follow and apply the advice offered by health workers and any other organisation working in the field of the effects of volcanic ash on human health; and to maintain the standpipes that will be made available to them so that they can no longer use water from the river as their main source of water.

PARTIAL CONCLUSION

In this part, which dealt with the methodological approach, presentation and discussion of the survey results, we wanted to get to grips with the real problem of the effects of volcanic ash on health in the Rusayo groupement, but also to verify our hypotheses through the different knowledge and perceptions of our respondents.

The aims of our survey were to test our hypotheses against the reality on the ground; to identify the effects on health of the volcanic ash emitted by the Nyiragongo volcano in the Rusayo groupement; and to find an effective, participatory solution to this problem.

To verify our hypotheses, we used descriptive, analytical and statistical methods based on documentation, questionnaires and free observation.

We found that the inhabitants of Rusayo have a good knowledge of volcanic ash, so our main hypothesis, which was that the inhabitants of Rusayo would have little knowledge of the harmful effects of the volcanic ash emitted by the Nyiragongo volcano on health, is invalidated, as shown in **Tables V, VI, X, XII and XIII.**

For the first specific hypothesis which was: The attitudes of the inhabitants of RUSAYO to the volcanic ash would be to panic and leave the fallout zone, to take it as a normal phenomenon; **Tables XIV**, **XXI** and **XXVII** show that 70.83% of the local population remain calm and hear the instructions while 88.89% of the health workers surveyed panic and leave the zone and 88.89% of the OVG researchers stay at home and hear the end of the fallout.

And for the second hypothesis which was: The practices of the inhabitants of RUSAYO to protect themselves from the volcanic ash would be : Move livestock away from the fallout zone, drink rainwater and do nothing special in the event of volcanic ash fallout; **Tables XV, XXII and XXVII** show that 52.08% of the local population surveyed 52.08% did nothing special, 88.89% of the health workers surveyed covered their nose and mouth once outside and 88.89% of the OVG researchers surveyed stayed at home and heard the end of the fallout.

As for the solution, 88.54% of the local population surveyed and 100% of the health workers surveyed advocate the construction of standpipes in the RUSAYO groupement, and 66.67% of the OVG researchers say that building the capacity of the inhabitants of RUSAYO in volcanic ash would be a solution to this problem **(Tables XVI, XXII and XXVIII).**

CHAPTER III

DEVELOPMENT PROJECT FOR THE CONSTRUCTION OF STANDPIPES IN THE RUSAYO DISTRICT

INTRODUCTION

In this section we will present a development project for the construction of standpipes in the Rusayo group, as proposed by most of our respondents; a project that will mitigate the risks of the effects of volcanic ash on health.

111.1.PROJECT IDENTIFICATION

111.1.1. Background and justification for the project

1. Project background

Around 600 million people worldwide live in areas potentially affected by volcanic hazards. During a volcanic crisis, people may be evacuated to protect them from life-threatening hazards (e.g. pyroclastic flows), but they may still be exposed to potentially dangerous atmospheric volcanic emissions. Volcanic ash is a ubiquitous hazard, potentially distributed over thousands of square kilometres. Inhaling ash can exacerbate existing asthma and bronchitis symptoms as well as respiratory symptoms. **(CJ Horwell at al. 2016)**

Volcanic eruptions affect the earth and humans around the world. When a volcano erupts and/or when there is volcanic activity in the central crater of a volcano, tonnes of volcanic ash are ejected into the atmosphere and deposited on the ground. The danger posed by volcanic ash is not limited to the area near the volcano, but can also affect a vast area. Ash ejected from the volcano affects people's daily lives, disrupting agricultural activities and damaging crops. **(Dian Fiantis at al. 2019)**

2. Project justification

Around 40,000 to 50,000 people living around the Nyiragongo volcano depend on rainwater collected in cisterns. This water is polluted by ash and lava needles (known as Pelee hairs) and becomes acidic **(Jacques Durieux at al.; 2011).**

Plumes of gas and ash from the Nyiragongo and Nyamulagira volcanoes are rapidly converted into acidic compounds, which cause local acid rain and air pollution, posing a potential danger to people, animals and agriculture and seriously damaging the environment. **(Cuoco et al. 2012)**

The RUSAYO group, a village located south-west of the Nyiragongo volcano, has no other source of drinking water apart from rainwater collected in cisterns. However, this water is affected by volcanic emissions, including volcanic ash, causing a number of illnesses among the area's inhabitants.

The construction of standpipes would be an effective solution to this problem, as they would be able to supply the entire Groupement with treated, well-preserved drinking water.

111.1.2. project objective

Reducing the risk of health effects from the volcanic ash emitted by the Nyiragongo volcano in the RUSAYO group.

111.1.3. The location and duration of the project

1. Project location

This project will be carried out in the Rusayo Group in the territory of NYIRAGONGO, in the province of NORD-KIVU in the Democratic Republic of Congo.

2. The duration of the project

Our project will take 12 months to build 21 standpipes, from January to December 2023.

111.1.4. The nature and legal framework of the project

1. The nature of the project

This project is of a social, health and environmental nature, as it will reduce the risks of the health effects of the volcanic ash emitted by the Nyiragongo volcano.

2. Legal framework of the project

The project involves a number of local and international organisations; in other words, it is a concerted effort.

111.1.5. PROJECT STAKEHOLDERS AND BENEFICIARIES

1. Project participants

To ensure the success of this project, it will be carried out in synergy, including :

- **Mercy corps and**
- **REGIDESO**

2. Project beneficiaries

- **Direct beneficiaries of the project:** The inhabitants of the Rusayo group are the direct beneficiaries of this project.
- **Indirect beneficiaries of the project:** the indirect beneficiaries are all the inhabitants of Nyiragongo territory.

111.1.6. Project strategies

To achieve the objectives of our project, we have proposed the following strategies:

- Mobilise financial resources;
- Contact leaders and authorities;
- Recruiting staff ;
- Site selection for the main reservoir and water treatment plant;
- Target areas for the 21 standpipes ;
- Construction of the main reservoir ;
- Construction of a water treatment plant ;

- Water catchments ;
- Construction of 21 standpipes;
- Water supply ;
- The handover of the standpipes to the Congolese government and
- Monitor project activities through the monitoring and evaluation system.

111.2.PROJECT STUDY

111.2.1. Timeliness of the project

In the course of our investigations, we noted a number of challenges faced by the inhabitants of Rusayo in relation to volcanic ash, including that of drinking rainwater, which is affected by volcanic emissions, causing a number of illnesses among the inhabitants. That's why this project is so timely, because it comes just when it's needed.

111. 2.2. Relevance of the project

This project is relevant because it comes from the population of the Rusayo group, which is the beneficiary of this project, which is why we consider it relevant.

112. 2.3 Project feasibility

a) From an economic point of view: The project is economically feasible because it will count first on local participation and then on the financial backer.

b) Financially: this project is financially feasible because its implementation is also made possible by the financial contribution of all those involved in the project.

c) Socially: this project is socially feasible because it solves a social problem.

d) On a technical level: this project will involve rural or community development coordinators, a secretary, an accountant, a logistician and various volunteers who can be found locally.

III.3. OPERATIONALISATION OF THE PROJECT

1. Organization chart

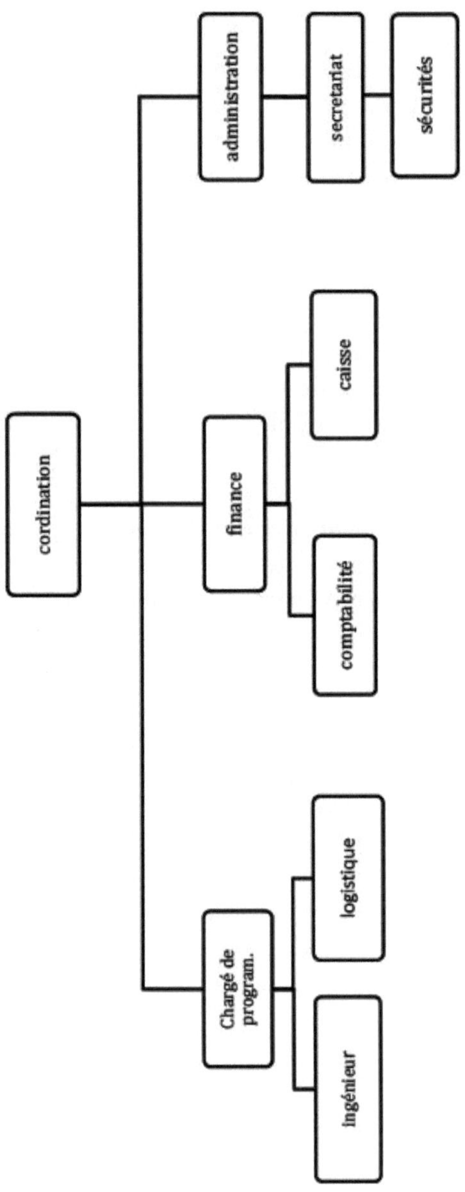

2. **How it works**

a) **Co-ordinator: the person who** organises all project activities.

b) **Programme manager:** responsible for the actual implementation of the project.

c) **Logistician:** all material resources are the responsibility of the logistician.

d) **Engineer:** responsible for supervising and coordinating standpipe construction activities;

e) **Accountant:** is in charge of the project's accounts and keeps an eye on the project's financial health.

f) **Cashier:** responsible for holding project assets and releasing them as and when required.

g) **Secretary: keeps** project documents up to date

111.4.INPUT PLANNING

1. Search for funds
2. Contacting leaders and authorities
3. Recruiting and training staff
4. Choosing the site for the main reservoir and water treatment plant;
5. Target areas for 21 standpipes ;
6. Buying construction materials;
7. Buying plumbing materials;
8. Building the main tank;
9. Building the water treatment plant;
10. Collecting water ;
11. Build 21 standpipes;
12. Water supply ;
13. Handing over the standpipes to the Congolese government;
14. Monitor project activities
15. Evaluating the project

Input 1: Mobilise the fund :

1. **Target:** Mercy Corps
2. **Objective: to** lease funds to the project
3. **Maturity:** from 01 January to 01 April 2023
4. **Duration:** 03 months
5. **Person responsible:** Coordinator
6. **Production:** Coordinator
7. **Resources**

- **Human resources:** Coordinator
- **Materials:** internet, computer, paper, pens
- **Financial:** local participation

Input 2: Contact leaders and authorities :

1. **Target:** leaders and authorities in Nyiragongo territory
2. **Objective: to** present civility and the project to leaders and authorities in Nyiragongo territory
3. **Maturity:** from 02 April to 04 April 2023
4. **Duration**: 3 days
5. **Person responsible:** Coordinator
6. **Production:** Coordinator
7. **Resource**

- **Human:** Coordinator
- **Equipment:** transport,
- **Financial :**

Input 3: Recruit staff:

1. **Target:** Nyiragongo Territory and the city of Goma
2. **Objective: To** provide the project with competent staff
3. **Maturity:** from 05 April to 05 July 2023
4. **Duration:** 3 months
5. **Person responsible:** Head of Personnel and Coordinator
6. **Director:** Head of Personnel
7. **Resource**

- **Human:** Coordinator
- **Equipment:** invoices, discharges.
- **Financial :**

8. **Sub-activities** :

III. Launch the offer
IV. Candidate selection
V. Text to selected candidates and
VI. The interview

Input 4: Choosing the site for the main reservoir and water treatment plant :

1. **Target:** the RUSAYO group
2. **Objective: to** secure land to house the main reservoir and water treatment plant
3. **Duration:** 1 month
4. **Deadline:** From 05 July to August 2023
5. **Person responsible:** Coordinator and programme manager
6. **Producer:** the programme manager and coordinator
7. **Resource**

- **Human:** the programme manager and the coordinator

- **Equipment:** plot occupancy
- **Financial :**

Input 5: Targeting environments for 21 hydrants :

1. **Target:** the RUSAYO group
2. **Objective:** find locations for the construction of standpipes
3. **Expiry date:** 06 August to 20 August 2023
4. **Duration:** 15 days
5. **Responsible:** the engineer and the programme manager
6. **Director:** the engineer and programme manager
7. **Resource**

- **Human:** the engineer and the programme manager
- **Equipment:**
- **financial:** transport costs, commission

Input 6: Purchase building materials :

1. **Target:** Large markets in the city of Goma
2. **Objective: To** have materials for construction
3. **Maturity:** From 21 August to 25 August 2023
4. **Duration:** 5 days
5. **Responsible:** programme manager, logistician and engineer
6. **Director:** logistician and engineer
7. **Resource**

- **Human resources:** logistician and engineer
- **Equipment:** cement, sand, concrete reinforcing bar, gravel, wheelbarrow, spades, etc.
- **Financial:** funds allocated to the business

Input 7: Purchase plumbing materials:

1. **Target:** Large markets in the city of Goma
2. **Objective: To** have plumbing materials
3. **Maturity:** From 25 August to 30 August 2023
4. **Duration:** 5 days
5. **Responsible:** programme manager, logistician and engineer
6. **Director:** logistician and engineer
7. **Resource**

- **Human:** logistician, engineer and plumber
- **Equipment:** pipes, taps, etc.
- **Financial:** funds allocated to the business

Input 8. build the main tank :

1. **Target:** the Rusayo cluster
2. **Objective: To** have a reservoir to store water

3. **Deadline:** From 01 September to 01 December
4. **Duration:** 3 months
5. **Responsible:** programme manager, logistician and engineer
6. **Director:** programme manager, logistician and engineer
7. **Resource**

- **Human resources:** programme manager, logistician, engineer, bricklayers and assistant bricklayers
- **Equipment:** cement, sand, concrete reinforcing bar, gravel, wheelbarrow, spades, etc.
- **Financial:** funds allocated to the business

Input 9: Building the water treatment plant :

1. **Target:** the Rusayo cluster
2. **Objective: To** have a water treatment plant
3. **Maturity:** from 10 September to 10 December 2023
4. **Duration:** 3 months
5. **Responsible:** programme manager, logistician and engineer
6. **Director:** programme manager, logistician and engineer
7. **Resource**

- **Human resources:** programme manager, logistician and engineer
- **Equipment:** filter pumps, cement, sand, concrete reinforcing bar, gravel, wheelbarrow, spades,
- **Financial:** funds allocated to the business

Input 10. Collecting water :

1. **Target:** Lake Kivu
2. **Objective:** find a sustainable source of water
3. **Maturity:** from 01 October to 01 December 2023
4. **Duration:** 2 months
5. **Responsible:** programme manager, logistician, engineer and plumber
6. **Director:** logistician, engineer and plumber
7. **Resource**

- **Human:** logistician, engineer and plumber
- **Equipment:** Pipes, cement, sand, concrete reinforcing bar, gravel, wheelbarrow, spades,
- **Financial:** funds allocated to the business

Input 11. Build 21 standpipes :

1. **Target:** the Rusayo cluster
2. **Objective: To** have standpipes
3. **Maturity:** from 05 October to 05 December 2023
4. **Duration:** 2 months

5. **Responsible:** programme manager, logistician and engineer
6. **Director:** logistician, engineer and plumber
7. **Resource**
- **Human:** logistician, engineer and plumber
- **Equipment:** Pipes, cement, sand, concrete reinforcing bar, gravel, wheelbarrow, spades,
- **Financial:** funds allocated to the business

Input 12. Water supply :

1. **Target:** the Rusayo cluster
2. **Objective: to** build a water distribution network
3. **Deadline:** From 10 to 13 December 2023
4. **Duration:** 4 days
5. **Responsible:** programme manager, engineer and plumber
6. **Director:** programme manager, the engineer and the plumber
7. **Resource**
- **Human:** programme manager, engineer and plumber
- **Equipment:**
- **Financial:** funds allocated to the business

Input13. Hand over the standpipes to the Congolese government:

1. **Target:** The Rusayo cluster
2. **Objective: to** make the project sustainable through participative management
3. **Expiry date:** 24 December 2023
4. **Duration:** 1 day
5. **Responsible:** Coordinator, programme manager and engineer
6. **Director:** Coordinator, programme manager and engineer
7. **Resource**
- **Human:** Coordinator, programme manager and engineer
- **Equipment:**
- **Financial :**

Input 14. Monitor project activities :

1. **Objective: to** monitor project activities on a daily basis
2. **Target:** All project staff
3. **Maturity:** From 01 January 2023 to 01 January 2024
4. **Duration:** 12 months
5. **Person responsible:** Coordinator and programme manager
6. **Director:** programme manager
7. **Resource**
- **Human:** Coordinator and Programme Manager

- **Material:** reports by department and monitoring sheet
- **Financial :**

Input 15: Evaluate project activities :

1. **Objective: To** assess the results of the project
2. **Target:** all project partners
3. **Maturity:** from 27 April to 01 May 2023

Duration: 5 days

From 27 to 31 August 2023

Duration: 5 days

From 28 December 2023 to 01 January 2024

Duration: 7 days

4. **Person responsible:** Donor and coordinator
5. **Director**: Financial backer and coordinator
6. **Resource**

- **Human:** all staff
- **Material:** monthly and quarterly reports, coordinator's report and project file
- **Financial :**

I. SCHEDULE OF ACTIVITIES

N0	^^^^JJeriods ActivïtèS^^^^	2023												2014
		January	**February**	**March**	**April**	**May**	**June**	**July**	**August**	**Seven.**	**October**	**Nov.**	**Dec.**	**January**
O1	Mobilising the fund													
02	Contacting leaders and authorities													
03	Recruiting and training staff													
04	Choosing the site for the main reservoir and water treatment plant													
05	Targeting environments for 21 standpipes													
06	Buying construction materials													
07	Buying plumbing materials													
08	Building the main tank													
09	Building the water treatment plant													
10	Collecting water													
11	Building 21 terminals													

	fountains													
12	Bringing water to Feau ;													
13	Handing over the standpipes to the Congolese government;													
14	Monitor project activities													
15	Evaluating the project													

Legend

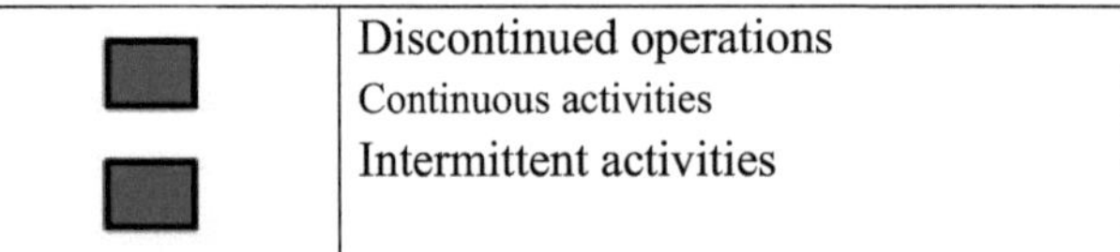

Discontinued operations
Continuous activities
Intermittent activities

II. PROJECT BUDGETING

Tableau 1. The staff

N°	Function	Grade	No.	Month	Monthly salary	Net salary
1	Coordinator	Laying off in Dvlpt	1	12	1000	12 000
2	Programme Manager	Upgrade to Dvlpt	1	9	900	8100
3	Finance Officer	Licensing in Accounts.	1	9	800	7200
4	Secretary	Graduating to Info	1	9	600	5400
5	Accountant	Graduate in Accounting	1	9	700	6300
6	Cashier	Graduate in Management	1	9	500	4500
7	Logistician	TDR	1	9	500	4500
8	Building engineer	Bac en batiment	1	5	900	4500
9	Plumber	Licensing plumbing	1	4	500	2000
	Construction labour					
10	Sentinel	Graduate	3	4	100	400
11	Bricklayers	Ir Building	30	4	520	2080
12	Mason's helpers	-	60	4	400	1600
TOTAL						58580$

Tableau 2. Purchase of construction equipment

N°	Designation	P.U in $ (in $)	Quantities	P.T. in $ (in $)
1	Cements	18	1000 bags	18 000
2	Concrete reinforcing bar	1500/T	1tonne	1500
3	Gravel	200/truck	4 Trucks	800
4	Sand	150/truck	10 trucks	1500
5	Stones	10$/m^3	300 m ^{3}	3000

TOTAL			24800

Tableau 3. Purchase of plumbing equipment

N°	Designation	P.U in $ (in $)	Quantities	P.T. in $ (in $)
1	**Tap**	6	30	180
2	**Pipe 10**	10	1000	10 000
3	**Pipe 50(2p)**	50	800	40 000
4	**Glue tan gite**	36	10 tins	360
4	**Filters**	1500	2 pieces	3000
TOTAL				53 540

Tableau 4. Rolling stock and fuel

N°	Designations	P.U	Quantities	Period	P.T in
1	**Trucks**	2000	2	3	12000
2	**Fuels**	1	4001/month	3	36 00
TOTAL					15 600

Table 4. Summary of budgeting

N°	Designation	Amount in
1	**Staff**	585 880
2	**Purchase of construction equipment**	24 800
3	**Purchase of plumbing equipment**	53540
4	**Rolling stock and fuel**	15 600
TOTAL		655 020

Table 5. Source of financing

N°	Partners	Contributions
1	**Mercy corps 60**	393 012$
2	**Local participation 40%.**	262 002$
TOTAL		655 020$

III. LOGICAL FRAMEWORK

^jgique hor. Green logic.	Narrative Summary (NS)	Objectively verifiable indicator (OVI)	Means of verification (MV)	Critical condition
Objective	Reducing the risk of health effects from the volcanic ash emitted by the Nyiragongo volcano in the RUSAYO group.	By 2024, the risk of health effects from the volcanic ash emitted by the Nyiragongo volcano in the RUSAYO group will be reduced.	• Consultants' report • Coordinator's report	-
Goal	The construction of 21 standpipes in the Rusayo group.	21 standpipes built by January 2024 in the Rusayo cluster.	• Coordinator's report • Programme manager's report • Videos, photos	That the construction of standpipes is well done
outputs	• Funds sought and obtained ; • Leaders and authorities contacted; • Staff recruited and trained ; • The land for the main reservoir and the water treatment plant chosen ; • The media for 21 targeted standpipes ; • Construction equipment purchased ; • Plumbing equipment	• 40% of funds granted by April 2023 and 60% during the project. • 10 Leaders and 18 local authorities contacted; • 9 agents recruited, 30 masons hired and 60 assistant masons hired and trained; • A plot of land for the main reservoir and water treatment plant chosen ; • 21 target environments for standpipes ;	• E-mail • Reloading, • Bills • Cheques • Extract from the bank • Work contrant • Receipts • Follow-up sheet • Photography • Rental contract • Employment contract	• That the research fund be obtained • That the site is a little far from households • That the office meets the criteria • Competent staff • That the equipment is of good quality • Make sure the hydrants are well
	purchased ; • The main reservoir built ; • The water treatment plant is building ; • Water catchments; • 21 standpipes built ; • The water supply is ; • The standpipes handed	• All construction materials purchased and stored • All plumbing materials purchased and stored ; • 1 reservoir built • 1 Teau treatment plant built ;		built • Regular monitoring • That the assessment is objective

	over to the Congolese government; • Project activities monitored ; • The project is assessed	• Several catchments of Teau made ; • 21 well-built hydrants; • The water supply is very well done throughout the Rusayo group; • 21 hydrants handed over to the Congolese government in good condition; • 48 follow-ups carried out • 4 assessments made		
Inputs	• Search for funds ; • Contacting leaders and authorities; • Recruiting and training staff; • Choosing the site for the main reservoir and water treatment plant;	• Staff salaries: $585,880 • Purchase of construction materials: $24,800 • Purchase of plumbing materials: $53540 • Rolling stock and fuel: $15,600 **Total: $655,020**	• Bills • Cheques • Staff pay slips • Cash book • Bank book • Discharge	• Ensure that funding arrives on time • The land is obtained at the scheduled time • That the equipment is purchased • That the constructions are carried out
	• Target areas for Zl hydrants ; • Buying construction materials; • Buying plumbing materials; • Building the main tank; • Building the water treatment plant; • Water catchment; • Build 21 standpipes; • Water supply ; • Hand over the standpipes to the			• Follow-up is ensured • That the project is evaluated

	Congolese government; • Monitor project activities; • Evaluate the project.			

GENERAL CONCLUSION

Here we are at the end of our work, which has been of capital importance to us because it has helped us to deepen our knowledge of the management of community problems by providing effective and participatory solutions; our work was on "**The knowledge, attitudes and practices of the inhabitants of Rusayo regarding the effects of volcanic ash on health**".

Our work consists of three chapters, the first of which presents the Rusayo group and general information on the knowledge, attitudes and practices of the inhabitants of Rusayo with regard to the effects of volcanic ash on health; the second chapter deals with the methodological approach, presentation and discussion of the survey results, and finally the third chapter includes a development project for the construction of standpipes in the Rusayo group.

To carry out this study we asked ourselves a series of questions, the **main one being**: What is the level of knowledge of the inhabitants of RUSAYO about the health effects of the volcanic ash emitted by the Nyiragongo volcano? And the **specific questions:** What are the attitudes of the inhabitants of RUSAYO towards the volcanic ash emitted by the Nyiragongo volcano? How do the people of RUSAYO protect themselves from the volcanic ash emitted by the Nyiragongo volcano? And finally, what is the solution for mitigating the health effects of the volcanic ash emitted by the Nyiragongo volcano?

To these questions we proposed some provisional answers by way of hypotheses; **Main hypothesis:** the inhabitants of RUSAYO would have little knowledge of the harmful effects of the volcanic ash emitted by the Nyiragongo volcano on health; and **Specific hypotheses: the** attitudes of the inhabitants of RUSAYO towards the volcanic ash would be to panic and leave the fallout zone, to take it as a normal phenomenon; the practices of the inhabitants of RUSAYO to protect themselves from the volcanic ash would be : to keep livestock away from the fallout zone, to drink rainwater and to do nothing special in the event of volcanic ash fallout; the construction of standpipes in the RUSAYO group and/or capacity-building for the people of RUSAYO on the subject of volcanic ash would be a solution to this problem.

At the start of this study, we set ourselves the following objectives:

Overall objective: To reduce the risk of health effects from the volcanic ash emitted by the Nyiragongo volcano in the RUSAYO groupement.

Specific objectives: To assess the attitudes of the inhabitants of RUSAYO to the volcanic ash emitted by the Nyiragongo volcano; To determine the practices of the inhabitants of RUSAYO with regard to the volcanic ash emitted by the Nyiragongo volcano; To develop a project to build standpipes in the RUSAYO group and/or to build the capacity of the inhabitants of RUSAYO with regard to the volcanic ash emitted by the Nyiragongo volcano.

To collect the data, we used a number of methods and techniques, including descriptive, analytical and statistical methods, as well as documentary techniques, questionnaires and free observation.

After surveys we noticed that the inhabitants have sufficient knowledge about volcanic ash, so our main hypothesis was invalidated. As for attitudes, they have good ones, so our first specific hypothesis was also invalidated. As for practices, they don't have any good ones, so our second hypothesis was confirmed. As for the solution to mitigate the effects on health of the volcanic ash emitted by the Nyiragongo volcano, the construction of standpipes was the best option for the inhabitants of Rusayo.

BIBLIOGRAPHY

A. Books and articles

1. **Annal of Ethiopia 2010/Vol 25**
2. **Cathy clerbaux at al.** *Measuring SO2 and volcanic ash with IASS, Meteorology, Weather and climate*, **2011**
3. **Charles M Balagizi at al.** *natural hazards in Goma and African rift system* **2018**
4. **Charles M Balagizi at al.** *Soil temperature and* CO_2 *degassing, fluxes and field observations before and after the febuary 29, 2016 new vent inside nyiragongo crater*,**2016**
5. **CUOCO E. at al.** *impact of volcanic plume emission on rain water chemistry the case of Mt. Nyiragongo in the Virunga volcanic region*, **2013**
6. **Dian Fiantis at al**. *Volcanic ash, insecurity for people but securing fertile soil for the future*, March 2019,p20
7. **Henry Gaudru**, *l'homme face aux risques volcaniques*, Paris, 2008 p180
8. **Horwell C I. at al.** *Physicochemical and toxicolocal profiling of ash from the 2010 and 2011 eruption of Eyjafjalla JOKULL and Crimsvotn Volcanos*, **2O12 p11**
9. **Horwell C I**. at al. Use of respiratory protection in Yogyakarta during the 2014 eruption of kelud, Indonesia community and agency perspective 2016
10. **Internation volcanic health hazard network (IVHHN), IAVCEI, GNS science and VSGS**; *The health risks of volcanic ash: a guide for the public*; 2003 P20
11. **The University of Florence, OVG and GENEVA;** *the health impact of volcanic eruptions,* **2005**
12. **LAUBET D**.B Jean louis; initiation aux méthodes de recherche en science sociales ; l'harmattan, Paris, p20
13. **International Civil Aviation Organization (ICAO),** *Volcanic Ash Manual, Radioactive Material and Toxic Chemical Cloud*, **2007 p80**
14. **Sarah SCAGLIONE at al**. *Environmental impact of volcanic emissions at nyiragongo(DRC),*2014 p19
15. **Teade NEIL**; *How a volcano in Iceland can disrupt European airspace,* 2010

B. TFC and Memories

1. **MAPENZI BALUME**, *knowledge, attitude of the population to the risks of volcanic activities on socio-economic life in the city of Goma, specific case of the commune of Goma, 2015*
2. **NDULU JUAKALI Médiatrice,** *Rôle de l'observatoire volcanologique de Goma sur la protection de la population de Goma, avant et après l4eruption du*

volcan nyiragongo du 17 janvier 2002 dans la ville de Goma, **2016**

3. **Julien LUKUBIKA,** *knowledge, attitudes and practices of households in the city of Goma on the risks associated with the vicinity of the Nyiragongo volcano, case of the Majengo neighbourhood,* **2019**

4. **Espoir B,** *The problem of the volcanic gas plume and its impact on the health of the population of Goma and the surrounding area*; Goma, 2013

C. Website

1. https://fr.wikipedia.org/wiki/Connaissance philosophy
2. www.grandlacksnews.comm/l-ovg-rassure-après-la-chute-des-cendres-volcanoes-in-the-zone-around-nyiragongo-volcano
3. https://fr.wikipedia.org/wiki/Connaissance philosophy
4. https://www.universalis.fr-encyclopedie-attitude-1 -the-concept-attitude
5. https://fr.wikipedia.org/wiki/Cendre volcano/definition
6. www.georisques.gouv.fr/dossier/volcanisme

Printed by Books on Demand GmbH, Norderstedt / Germany